THE LIBRARY OF THE PLANETS™

MERCURY

Amy Margaret

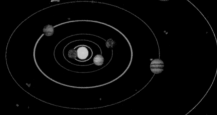

The Rosen Publishing Group's
PowerKids Press™
New York

For sweet Hannah and precious Talia

Published in 2001 by The Rosen Publishing Group, Inc.
29 East 21st Street, New York, NY 10010

First Edition

Book Design: Michael Caroleo and Michael de Guzman

Photo Credits: pp. 1, 4 PhotoDisc; p. 7 (Roman god Mercury) Michael R. Whalen/NGS Image Collection, p. 7 (Mercury) PhotoDisc (digital illustration by Michael de Guzman); p. 8 Courtesy of NASA/JPL/California Institute of Technology; p. 11 PhotoDisc (digital illustration by Michael de Guzman); p. 12 (core) Davis Meltzer/NGS Image Collection, p. 12 (inner planets) PhotoDisc (digital illustration by Michael de Guzman); p. 15 Chris Foss/NGS Image Collection; p. 16 (Discovery scarp) Courtesy of NASA/JPL/California Institute of Technology; p. 19 © Dewey Vanderhoff/Liaison Agency; p. 20 (Mercury) PhotoDisc, p. 20 (*Mariner 10*) Courtesy of NASA/JPL/California Institute of Technology (digital illustration by Michael de Guzman).

Margaret, Amy.
    Mercury/ by Amy Margaret.
        p. cm.– (The library of the planets)
    Includes index.
    Summary: Describes the history, unique features, and exploration of Mercury, the
planet closest to the Sun.
    ISBN 0-8239-5642-3 (library binding : alk. paper)
    1. Mercury (Planet)–Juvenile literature. [1. Mercury (Planet)] I. Title. II. Series.

QB611 .M28 2000
523.4–dc21   99-040235

Manufactured in the United States of America

Contents

The Planet Mercury

Mercury is the second smallest **planet** in our **solar system**. There are nine planets in our solar system. Each moves around the Sun, a star that gives off heat and light. Of all the nine planets, Mercury is the closest to the Sun.

Our solar system seems very large, but it is really one small part of the **galaxy** in which we live. A galaxy is a very large group of stars. Our galaxy is called the Milky Way. There are many galaxies beyond the Milky Way.

Like the other planets in our solar system, Mercury may have been created as a result of a huge explosion called the Big Bang. Many scientists believe that the Big Bang occurred about 10 billion years ago. This explosion spread gases, which cooled and then formed galaxies. Our solar system, including Mercury, could have been made about four and a half billion years after the Big Bang.

Mercury is the second smallest planet in our solar system.

The planet Mercury got its name from Roman **mythology**. There is a story about a messenger named Mercury who the Romans believed took messages to their gods. Mercury the messenger moved very quickly, just like the planet does. Mercury moves around the Sun at almost 30 miles (48.2 km) a second. That is faster than any other planet in our solar system. When ancient Greek **astronomers** first observed the planet, they thought it was two separate stars instead of just one planet. In the early morning light, the astronomers saw the planet on one side of the Sun. During the evening they saw the faint light of Mercury on the other side of the Sun. This was because Mercury moved quickly around the Sun. In about 350 B.C., a Greek astronomer, Heraclides of Pontus, figured out that the two stars that had been observed were actually one planet, Mercury.

Mercury is named after the messenger who delivered messages to Roman gods.

Exploring Mercury

Mercury has been explored in many ways. The earliest astronomers looked up in the sky to observe the tiny planet. In the mid-seventeenth century, the first **telescopes** were built so people could see more details in space.

Astronomers throughout the years tried to use telescopes to study Mercury. Mercury's gray and black **surface** was hard to see because it was close to the Sun and the Sun's bright light blocked it. Unfortunately, telescopes were not able to help scientists gather much information about Mercury. Beginning in the late 1950s, **scientists** used **space probes** to see the planets up close. A space probe is a spacecraft that does not carry any people. Scientists on the ground use computers to steer the probe. The space probe sends back photographs to Earth. In 1973, a space probe was sent to observe Mercury for the first time. This space probe was called *Mariner 10*.

The space probe Mariner 10 is the photo on top. The photo of Mercury on the bottom was taken by Mariner 10.

How Mercury and the Planets Move

Each planet in our solar system moves in two different ways. First, each planet spins on its **axis**, like a merry-go-round or a top. Second, the planets circle the Sun.

All nine planets in our solar system **orbit**, or move around, the Sun at different speeds. Some take less than one Earth year, while other planets spend many, many Earth years to complete one trip. By looking at the chart on the next page, it is easy to see that Mercury orbits the Sun the fastest. Find out how far the other planets are from the Sun and how long it takes these planets to travel around it. Notice that the farther the planet is from the Sun, the longer it takes to move around it.

The drawing at the top shows Mercury orbiting the Sun. This takes 88 Earth days. Mercury takes about 59 Earth days to spin on its own axis.

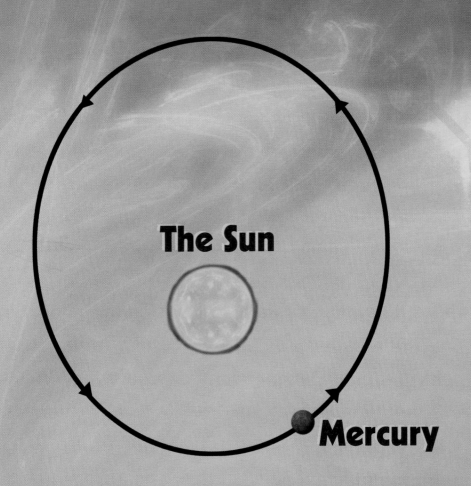

The Sun

Mercury

FUN FACTS

Planet	Orbit Time Around the Sun
Mercury	88 Earth days
Venus	225 Earth days
Earth	1 year (365 days)
Mars	1 year and 322 Earth days
Jupiter	12 Earth years
Saturn	29 1/2 Earth years
Uranus	84 Earth years
Neptune	165 Earth years
Pluto	248 1/2 Earth years

Crust

Mantle

Core

Special Features of Mercury

Mercury is one of the four "inner planets." The other three are Venus, Earth, and Mars. They are called the inner planets because they are the closest planets to the Sun. The inner planets are each made up of three parts. These parts are the surface, called a crust, a middle layer known as the mantle, and the core, or center of the planet. Mercury has one of the largest cores. It is made up of iron, which makes it a very heavy planet.

Mercury has a wider range of **temperatures** on its surface than any other planet. The side of Mercury facing away from the Sun can get as cold as -300 degrees Fahrenheit (-184.4 degrees C). When it faces the Sun, the temperature can rise all the way to 800 degrees Fahrenheit (427 degrees C). The coldest temperature recorded on Earth is -100 degrees Fahrenheit (-73.3 degrees C), and the hottest temperature recorded is close to 130 degrees Fahrenheit (54.4 degrees C).

The picture on top shows the crust, mantle, and core of the inner planets. The picture on the bottom is of the four inner planets. Mercury is the small planet closest to the sun.

Mercury's Craters

Mercury's surface is marked with **craters**, like the Moon's surface. A crater is a large depression or hole. It can be created when something large, like a **meteorite**, hits the surface. Scientists think that falling **comets**, **asteroids**, and meteorites have made Mercury's craters. The Moon has many craters, and some are on top of each other. On Mercury the craters are more spread out and have smooth plains in between them.

The Caloris Basin is the largest known crater on Mercury. It is also one of the largest in the whole solar system. It was formed about four billion years ago when a large asteroid hit Mercury. It was named Caloris,

Mercury looks a lot like the moon. Both have many craters, but the craters on Mercury are more spread out.

the Latin word for "heat." The name fits well because
this spot has the highest temperatures on the
planet. This area faces the Sun when
Mercury is closest to
the Sun.

Discovery Scarp

The Scarps of Mercury

Mercury's surface also has cliffs, or **scarps**. From a space probe, scarps look like giant wrinkles. Each scarp is one to two miles (1.6 to 3.2 km) high and can run for hundreds of miles (km) along the surface of a planet. A scarp is formed when an object, such as a meteorite, hits the planet's surface. The crash causes the land around the meteorite to move outward in waves, creating scarps. (Think of what happens when you drop a rock in a pool of water.) One of the largest known scarps on the planet Mercury is the Discovery scarp. It is 310 miles (499 km) long and almost 2 miles (3.2 km) high. It was found through pictures taken by the space probe, *Mariner 10*, in 1974 and 1975. Some of the scarps on Mercury have been named after ships used for early exploration of the seas. The Discovery scarp was named after one of the ships used by English explorer James Cook, who sailed the Pacific Ocean in the late 1770s. Cook's ship was called the *Discovery*.

The Discovery scarp cuts across several craters. Scientists think that the scarp was made after the craters had already been formed.

Seeing Mercury From Earth

Of all the planets we can see from Earth without a telescope, Mercury is the most difficult to find. Mercury is very close to the Sun, so the Sun's bright light keeps us from seeing the planet clearly. If you want to see Mercury, there are two good times to give it a try.

Look for Mercury in the eastern sky one hour before the Sun rises. You can also wait until just after the Sun sets and look for it

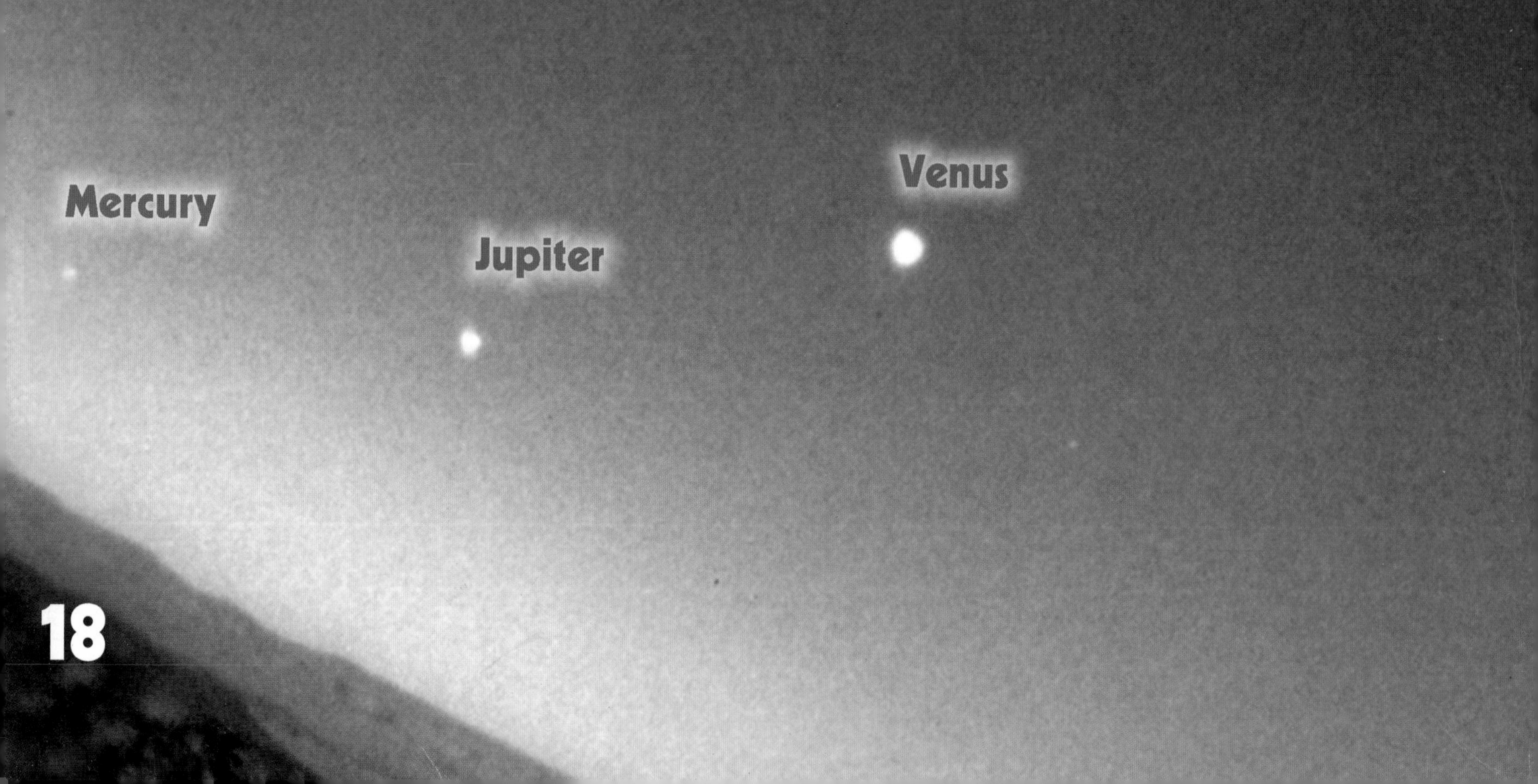

Mercury

Jupiter

Venus

in the western sky. Mercury is a very bright planet and can outshine many stars in the sky. To find the right part of the sky, remember that the Sun rises in the east and sets in the west.

FUN FACTS

For the latest on night sky planet sightings, look up this internet address. It will give you the most up-to-date, best times of the year to see Mercury.
http://www.kidsnspace.org/what_can_i_see.htm

If you weigh 100 lbs. (45.4 kg) on Earth, you would weigh 38 lbs. (17.2 kg) on Mercury.

The photograph on the left is of Mercury, Jupiter, and Venus. It was taken from high above Earth's atmosphere.

Our First Mission to Mercury

The *Mariner 10* was a space probe launched on November 3, 1973. It took the *Mariner 10* five months to travel to Mercury, a distance of almost 60 million miles (96.5 million km)! *Mariner 10* continued around the Sun, passing by Mercury two more times during its journey.

Scientists knew that the space probe would be traveling very close to the Sun. It had to be protected from the Sun's heat. A sunshade was built for the *Mariner 10* by placing special layered blankets on the probe's top and bottom.

The space probe had two cameras that photographed close to half of Mercury's surface. Computers on the spacecraft then sent the pictures to Earth. *Mariner 10* showed us how similar Mercury's surface is to our Moon's surface. Both are colorless in appearance and have many craters.

This picture shows a photograph of Mercury taken by Mariner 10. At the bottom right is a picture of the space probe itself.

Future Missions to Mercury

No space probes have been sent to Mercury since the *Mariner 10* mission in 1973. Space groups, such as NASA (the National Aeronautics Space Administration), are working on new missions to further explore the planet. The next mission's space probe will include an instrument that looks for water and a special camera that can map ice below the surface. One of the aims will be to take pictures of those areas that the *Mariner 10* did not photograph.

Other groups are also planning to send missions to Mercury. The Carnegie Institution of Washington, D. C., is sending a spacecraft called *Messenger* to enter Mercury's **atmosphere** in September of 2009. It will return one year later with new research. Future missions will try to gather more information about Mercury's surface, crust, atmosphere, and varying temperatures. By understanding this mysterious inner planet, scientists hope to gain more knowledge about our solar system.

Glossary

asteroids (as-tehr-OYDZ) Small objects in space that revolve around the sun.

astronomers (ah-STRAH-nuh-merz) Scientists who study the night sky, the planets, moons, stars, and other objects found in space.

atmosphere (AT-muh-sfeer) The layer of gases that surrounds an object in space. On Earth, this layer is the air.

axis (AK-sis) A straight line on which an object turns or seems to turn.

comets (KAH-mits) Heavenly bodies, made up of ice and dust, that look like stars with tails of light.

craters (KRAY-terz) Holes in the ground that are shaped like a bowl.

galaxy (GAH-lik-see) A large group of stars and the planets that circle them.

meteorite (MEE-tee-or-yt) A rock that has reached a planet from outer space.

mythology (mih-THOL-oh-jee) Stories that people make up to explain events in nature or history.

orbit (OR-bit) To circle around something.

planet (PLAN-et) A large object, like Earth, that moves around the sun.

scarps (SKARPS) Cliffs formed when an object, such as a meteorite, hits a planet's surface.

scientists (SY-en-tists) People who study the way things are and how things act in the world and universe.

solar system (SOH-ler SIS-tem) A group of planets that circles a star. Our solar system has nine planets that circle the sun.

space probes (SPAYS PROHBZ) Spacecrafts that travel in space and are steered by scientists on the ground.

surface (SER-fis) The top or outer covering of something.

telescopes (TEL-uh-skohps) Instruments used to make distant objects appear closer and larger.

temperatures (TEMP-pruh-cherz) How hot or cold things are.

Index

A
asteroids, 14
astronomers, 6, 9
atmosphere, 22
axis, 10

B
Big Bang, 5

C
Caloris Basin, 14
comets, 14
Cook, James, 17
core, 13
craters, 14, 21
crust, 13, 22

G
galaxy, 5

H
Heraclides of Pontus, 6

M
mantle, 13
Mercury, the messenger, 6
meteorite, 14, 17
Milky Way, 5
mythology, 6

O
orbit, 10

S
scarps, 17
scientists, 9, 14, 22
space probes, 9, 17, 21, 22
solar system, 5, 6, 10, 22
surface, 9, 13, 22

T
telescopes, 9, 18
temperatures, 13, 15, 22

If you would like to learn more about Mercury, check out this Web site:
http://www.dustbunny.com/afk/

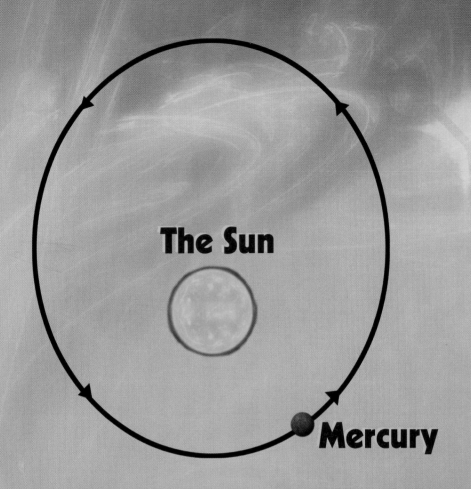

The Sun

Mercury

FUN FACTS

Planet	Orbit Time Around the Sun
Mercury	88 Earth days
Venus	225 Earth days
Earth	1 year (365 days)
Mars	1 year and 322 Earth days
Jupiter	12 Earth years
Saturn	29 1/2 Earth years
Uranus	84 Earth years
Neptune	165 Earth years
Pluto	248 1/2 Earth years

Crust

Mantle

Core

THE LIBRARY OF THE
PLANETS™

MERCURY

Amy Margaret

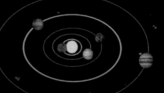

The Rosen Publishing Group's
PowerKids Press™
New York

For sweet Hannah and precious Talia

Published in 2001 by The Rosen Publishing Group, Inc.
29 East 21st Street, New York, NY 10010

First Edition

Book Design: Michael Caroleo and Michael de Guzman

Photo Credits: pp. 1, 4 PhotoDisc; p. 7 (Roman god Mercury) Michael R. Whalen/NGS Image Collection, p. 7 (Mercury) PhotoDisc (digital illustration by Michael de Guzman); p. 8 Courtesy of NASA/JPL/California Institute of Technology; p. 11 PhotoDisc (digital illustration by Michael de Guzman); p. 12 (core) Davis Meltzer/NGS Image Collection, p. 12 (inner planets) PhotoDisc (digital illustration by Michael de Guzman); p. 15 Chris Foss/NGS Image Collection; p. 16 (Discovery scarp) Courtesy of NASA/JPL/California Institute of Technology; p. 19 © Dewey Vanderhoff/Liaison Agency; p. 20 (Mercury) PhotoDisc, p. 20 (*Mariner 10*) Courtesy of NASA/JPL/California Institute of Technology (digital illustration by Michael de Guzman).

Margaret, Amy.
    Mercury/ by Amy Margaret.
        p. cm.– (The library of the planets)
    Includes index.
    Summary: Describes the history, unique features, and exploration of Mercury, the
planet closest to the Sun.
    ISBN 0-8239-5642-3 (library binding : alk. paper)
    1. Mercury (Planet)–Juvenile literature. [1. Mercury (Planet)] I. Title. II. Series.

QB611 .M28 2000
523.4–dc21   99-040235

Manufactured in the United States of America